Forensic Light: A Beginner's Guide

* * * *

David Rudd Cycleback

Forensic Light : A beginner's Guide
 by David Rudd Cycleback

Publisher: Hamerweit Books
ISBN: 978-0-578-02906-1

Contents

(1)
Introductory Notes

This small book is a beginner's guide to ultraviolet, visible and infrared light and their practical uses. These forms of so-called *forensic light* have long been used in a wide variety of areas. This includes authenticating currency and licenses, identifying forged and altered art and collectibles, purifying air and water, dating paper, investigating crime scenes, identifying and curing medical conditions and making glow in the dark art and crafts. The list goes on and on.

While a primer hardly intends to cover it all, even the beginner reading this book can learn how to do things like identify modern reprints of antique posters and sports cards, and make invisible security markers for valuable heirlooms.

Ultraviolet and infrared light are invisible to human eyes. Despite being invisible, ultraviolet is useful because it makes some materials fluoresce, or give off visible light. The color and intensity of this fluorescence helps identify and authenticate material. For example, antique Burmese art glass tends to fluoresce a bright yellow under black light, while modern reproductions usually do not. The current US$10 bill has a red band under black light, while the US$20 has a green band. An 1860s celebrity photo or theatre program that fluoresces bright light blue is a modern reprint.

Infrared doesn't make things visibly fluoresce but with an infrared viewer or camera we can see details and qualities hidden in daylight. Some US currency has security stripes that can only be seen with an infrared viewer. Art historians use infrared viewers and cameras to view the original sketches and paintings behind the surface layer of paint on a painting. This is an essential part of determining if that museum painting was really painted by Rembrandt.

Forensic visible light involves things you can see with your naked eyes but don't normally pay close attention to. For example, holding genuine and reprint 1975 Topps baseball cards up to a desk lamp will reveal different opacities, meaning more light shines through one card than the other. This helps show that the cards are made of different cardstock. Further inspection may show that one card's front surface reflects much more light (is glossier) than the other. These and similar visual light tests are simple but effective in counterfeit and reprint detection.

In the areas of authentication and fake/forgery detection there are two main reasons for learning how to use forensic light. The first and most obvious is forensic light helps you better identify genuine, fake, reprinted and altered items.

The second is because in legal dispute, scientific debate or other formal setting, objective scientific documentation is often essential to proving the case. It is one thing for a so-called expert witness to simply say the postcard in question 'looks like a reprint.' "Why should I take your word for it?," the arbiter unfamiliar with postcard authentication might say. It's another thing for the expert to say the card appears to be a reprint, while providing infrared pictures, ultraviolet readings and other scientific data to help back up his conclusion. Not only will this help convince the arbiter, judge and/or jury, but the scientific documentation can be verified by other experts. If you tell the judge a '1910 poster' can be identified as a modern reprint in part as it fluoresces bright light blue under black light, another expert may tell the judge that your black light interpretation is correct.

Further, documenting your forensic light research on toys or prints or pins or whatever you collect can be useful not only for you, but fellow collectors and dealers. Documenting the glossiness, opacity and/or black light colors of a set of popularly collected trading cards can be information used by others to judge the authenticity of their items.

The reader is encouraged to apply forensic light to his or her areas of interest, to discover the trends and limits. This book was written for a course, with the final project allowing the student to report on his findings and research in his area of interest.

Forensic light and this book are a supplement to your other knowledge, hands on experience, resources and tools. This other knowledge includes familiarity and knowledge in your specialty (baseball cards, postcards, antique photographs, gems), and asking for input from fellow hobbyists and experts. Many readers are already seasoned in their area of collecting or dealing and want to learn about forensic light to make their opinions even more assured.

Forensic light and science in general are helpful to authentication but have limits. Science can identify many fakes, forgeries and reprints. However, authentication requires additional information and thought. A 'Teddy Roosevelt autographed letter' can be proven to be a forgery when a black light shows the paper to have been made long after Roosevelt died. However, if scientific tests show the paper is vintage, that doesn't by itself prove the letter authentic. A modern forger can find vintage paper to write on. That the paper is shown to be vintage is helpful to the authentication process, but authentication requires other testing and judgment including analysis of the handwriting, looking at the provenance, perhaps getting opinions from one or more outside signature experts.

Forensic light is important for collectors, dealers, historians and auctioneers, but is still just one of many tools to be used.

Forensic light equipment used in this book

Black light (also known as longwave ultraviolet light, UVA light)

Shortwave ultraviolet light, aka UVC (some lights give off both longwave and shortwave black light)

Infrared digital camera or infrared viewer

The reader is not required to buy all equipment, and many will stick with the inexpensive black light. Even if the reader eventually accumulates all the equipment, he may start with a black light, then later on down the road buy an infrared camera or shortwave ultraviolet light. The only expensive item is the infrared viewer/camera. The other items can be purchased cheaply.

Safety of the equipment
The black light (UVA) is safe if used correctly and the infrared viewer and infrared camera are safe. The UVC shortwave ultraviolet light is the most dangerous, but safe if used correctly and prudently. More details on safety are described later in the book.

(2)
A Brief Overview of Ultraviolet Light

Ultraviolet, often called UV, is a form of light invisible to humans. UV makes up a small section on the entire spectrum of light. As shown below, the entire spectrum also includes x-rays, gamma rays and the visible light we see.

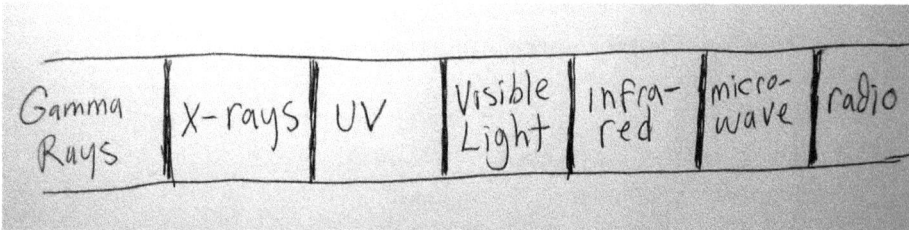

Light is commonly categorized by its wavelength. The light to the left in the above picture has the shorter wavelength, while the light on the right side has the longer. Ultraviolet light is just to the left (shorter) of the color violet in the visual light spectrum. Infrared is just to the right (longer) of the color red in the visual light spectrum. Humans can only see visible light and its colors violet to red. Human can't see any of the other type of light. The visible colors have different wavelengths from each other. Blue has a shorter wavelength than green, orange has a shorter wavelength than red, and so on.

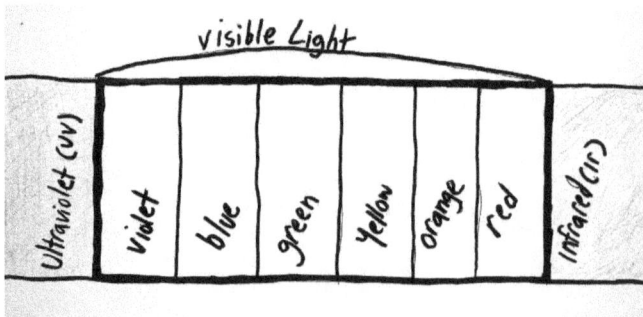

The Different Categories of UV Light, Including Black Light

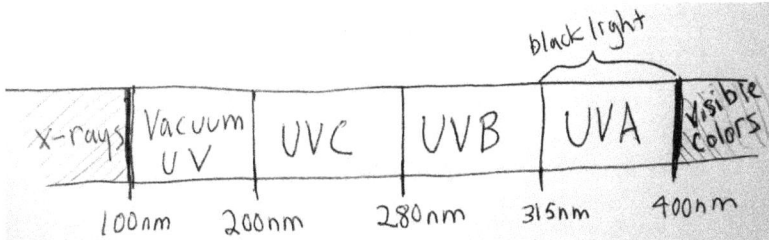

Ultraviolet light itself is commonly divided into categories. As with all light, the sections are defined by the wavelength.

The most common categories you will see are **UVA** (also known as longwave UV and black light), **UVB** (a.k.a mid-wave UV) and **UVC** (a.k.a. shortwave UV and germicidal light).

Common names:
UVA = longwave UV = black light
UVB = midwave UV
UVC = shortwave UV = germicidal light (kills germs)

Measuring and representing ultraviolet light: wavelengths
Light is represented and measured in different ways. The most common way is as waves. Length of the wave is measured from crest to crest, though you can also measure it bottom to bottom.

Wavelength is commonly measured in **nanometers (abbreviated as nm)**. A nanometer is one billionth of a meter, or one millionth of a millimeter. A human hair is roughly 30,000 nm thick. One will occasionally find light represented in **Angstroms (A)**, which is one tenth (1/10th) of a nanometer. An easy conversion.

1 nanometer = 10 Angstroms
1 Angstrom = 0.1 nanometer

The following are the wavelengths in nanometers for UVA, UVB and UVC:

UVA (longwave, black light) = 320 to 400 nanometers (nm)
UVB (mid-wave) = 280-315nm
UVC (shortwave, germicidal) = 200-280 nm

Most black lights, including the one you use for this book, are in the 380s-390s nm range. This is just a tad longer in wavelength than visible violet light.

Most germicidal lamps are 254nm. Germicidal light is dangerous for human skin and eyes, but is stopped by ordinary glass and even clothing.

(3)
Your Tool : a Black Light

Black light = UVA = longwave UV = 320-400 nm

Along with discussing all kinds of ultraviolet and their uses, this book shows you how to use a black light, also known as a longwave UV or UVA light. This chapter is about how to buy the correct kind of black light.

As discussed in previous chapter, there are different kinds of ultraviolet light. The most common kinds are UVA, UVB and UVC. The type of ultraviolet light you want to purchase for this part is the UVA, or longwave ultraviolet light. These are commonly marketed and referred to by the nickname "black light." When purchasing a black light, make sure it is the safe UVA kind, and not the more dangerous UVC kind (Use of UVC is discussed in a later chapter).

Luckily, the longwave ultraviolet light or black light is the most common and inexpensive ultraviolet light on the market.

Buying your black light.
There are many places to buy a good black light. You can pick up cheap black lights from amazon.com and ebay.com. Some science, hobby and rock shops sell them.

Black lights come in many styles and powers. This includes screw in bulbs and large and small flashlights. I own a small flashlight style and a screw in bulb. Both were inexpensive and serve different purposes. The bulb screws into a standard light socket and the flashlight can be carried around in my pocket when I leave home.

As long as the light gives off black light, the style you use is up to you.

The little flashlight is good for authenticating art, currency and such. It is portable and can be carried around most anywhere.

LED and other high powered black lights are good for rock hunting and general inspection, and are also good for examining art and such.

Screw in light bulbs are especially good for glow in the dark art and crafts decorations like posters, paintings and clothes. This type of light can also work well for inspecting art and currency, but isn't as portably convenient as a flashlight.

Be careful when purchasing screw in black lights, as many are not ultraviolet. Make sure it specifically mentions that it is UV or makes things glow in the dark. Most UV light bulbs are fluorescent, typically with a 'curly cue' bulb design (see picture on next page). Many incandescent bulbs are not ultraviolet, and are the old time oval bulb shape. The non-UV incandescent light bulbs are cheap, so if you pick the wrong one you will only be out a couple of bucks.

Examples of commonly found black light styles:

Popular hand held black light. Runs of AA batteries.

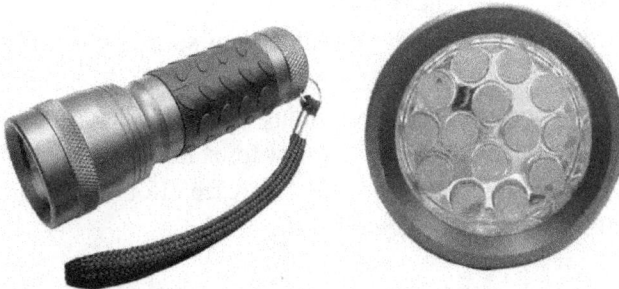

Miniature black light flashlight that uses Light-Emitting Diodes (LEDs). These often give off lots of light and are easily portable.

Fluorescent black light bulb with the distinct curly cue design. This screws into a normal light socket, like on a living room lamp.

(4)
How to Use Your Black Light

Once it's plugged in or the batteries popped in, most black lights are as easy to use as normal flashlights. The black light is used in the dark, the darker the better. They can work outside at night and inside in a dark room. You should first stay in the dark for at least a couple of minutes so your eyes get adjusted to the dark. After that, shine the black light around and you should find things that *fluoresce,* meaning they glow the in dark. Most black lights emit a small amount of visible light so that you know it's on.

When you are later examining specific objects—like a baseball card or dollar bill—it's best to examine your specimen against something that does not fluoresce. If the background fluoresces brightly this may affect the results.

* * * *

Safety of black light
You'll be happy to know that UVA/ black light is the safest type of the ultraviolet light. The light you will use is just higher in frequency than visible light. In fact, sunlight and office lights contain UVA light, so you're exposed to it on a daily basis. It is UVC, or shortwave, that is more dangerous and extra care is to be taken.

While black light is not of great danger, reasonable care should still be taken. The key with black light is to not stare directly at the light source, just as you shouldn't stare directly at the sun or a regular light bulb. And, as with sunlight, don't overdo exposure. Don't try and suntan with your black light. If you want, you can wear a strong UVA/UVB protecting suntan lotion, just as you should be wearing outside in the sunlight.

* * * *

Test your black light around the house

In the dark, go around your home or office and see which things fluoresce and which do not. Common around the house things that fluoresce include:

White paper
Some cloth, including parts of shirts, hats
Laundry detergent
Eyeglasses
Tennis balls
Some glass and plastics

Some things fluoresce so brightly you can almost read by the light!

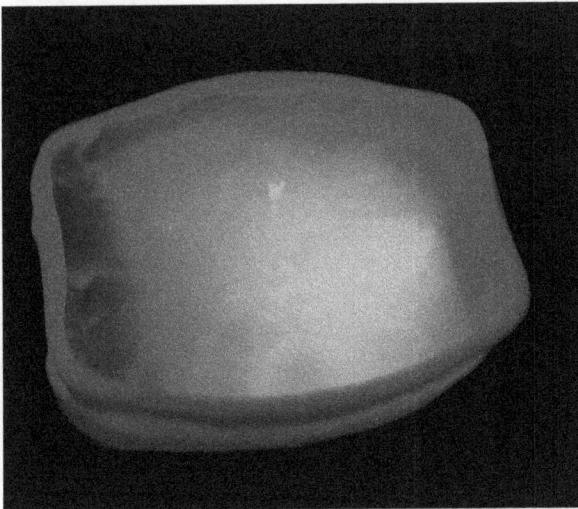
petroleum jelly in jar

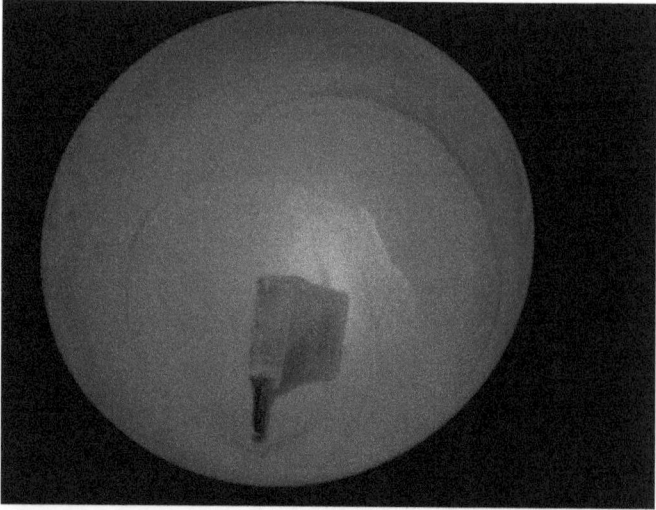

laundry detergent bucket

white thread on spool

(5)
How Does Black Light Make Things Fluoresce?

The fluorescence, or visible light that is emitted from a material when black light is shined on it, happens at the atomic level. When you are shining a black light on an object, you are testing the object's atomic makeup.

Just as with light, heat and x-rays, black light is a form of energy. When black light is shined on a material, whether the material is glass, plastic or paper, energy is being added to the atoms of the material. The atoms can only hold this extra energy for a short amount of time before having to release it. The atoms give off the energy in a different form than received. The atoms receive the energy as black light, but may give the energy off as heat, ultraviolet light, infrared light, visible light or, often, a combination of these. What form(s) of energy the atoms gives off is dependant on the makeup of the atoms.

If visible light is emitted by the atoms, that is the fluorescence we see. The color of this visible light is also dependant on the atomic make up. If the atom gives off just heat, ultraviolet light or infrared light, there will be no fluorescence. In a darkened room this material will remain dark.

Phosphorescence : After glow
Phosphorescence is closely related to fluorescence. Like fluorescent materials, phosphorescent materials give off visible light when excited by energy like UV light. However, while fluorescent material quits emitting light when UV light is turned off, phosphorescent material continues to give off light. The extra duration varies from phosphorescent material to

phosphorescent material. Some phosphorescent material gives off light for a fraction of a second longer, other material for hours or even days. The phosphorescence can be caused by UV light, but also visible light, X-rays, infrared and other light. What kind of light(s) produces the phosphorescence depends on the material. As with fluorescence, the color, brightness and duration of the phosphorescence are caused by the atomic make up of the material.

As with fluorescent material, the added energy of UV or other light excites the atoms in phosphorescent material, raising the electrons to a higher orbital. While the electrons move back to their normal orbital right away with fluorescent material, it takes longer with phosphorescent material. Thus the phosphorescent glow lasts longer.

Did you know?

Ultraviolet astronomy is a part of astronomy that observes and studies the ultraviolet light given off by stars, planets, galaxies and the cosmos. As much of this light is beyond UVA and UVB and blocked by the earth's atmosphere, the light can only be observed from space or the upper atmosphere. UV telescopes are carried by rockets, the Hubble Space Station and the Space Shuttle. From the UV light, scientists can deduce the chemical makeup, weight, temperatures and even age of the cosmic bodies.

(6)
Black Light: Identifying Modern Reprints and Fakes of Antique Paper Items

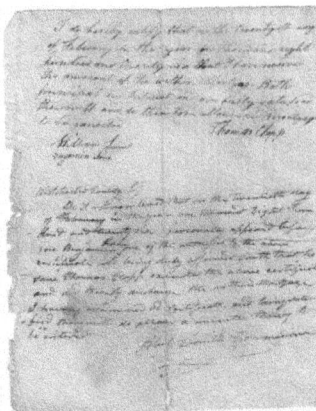

A black light is effective in identifying many, though not all, modern paper stocks. This allows the collector and dealer to identify modern reprints and forgeries of antique trading cards, posters, photographs, programs and other paper memorabilia. Many collectors buy a black light specifically for this purpose.

Starting in the late 1940s, manufacturers of many products began adding optical brighteners and other new chemical compounds to their products. Optical brighteners are invisible dyes that fluoresce brightly under ultraviolet light. They were used to make products appear brighter in normal daylight, which contains some ultraviolet light. Optical brighteners were added to laundry detergent and clothes to help drown out stains and to give the often advertised *whiter than white whites*. Optical brighteners were added to plastic toys to makes them brighter and more colorful. Paper manufacturers joined the act as well, adding optical brighteners to many, though not all of their white papers stocks.

A black light can identify many trading cards, posters, photos and other paper items that contain optical brighteners. In a dark room and under black light optical brighteners will usually fluoresce a very bright light blue or bright white. To find out what this looks like shine a recently made white trading card, snapshot or most types of today's printing paper under a black light. If paper stock fluoresces very bright as just described, it almost certainly was made after the mid 1940s. It is important to note that not all modern papers will fluoresce this way as optical brighteners are not added to all modern paper. For example, many modern wirephotos have no optical brighteners. This means that if a paper doesn't fluoresce brightly this does not mean it is necessarily old. However, with few exceptions, if a paper object fluoresces very brightly, it is modern.

The great thing with this simple test is you don't have to be an expert in an area to identify modern fakes and reprints. You may know next to nothing about silent era Hollywood movie posters or 1890s fishing industry pamphlets, but you can still identify many reprints and fakes.

(7)
Black Light: Art Glass

Black light is a useful tool for judging the identity and age of art glass vases, figurines and more. Different types and ages of glass can fluoresce different colors, and the color of fluorescence can be helpful in identification. As there are variations and exceptions, the fluorescent colors should be used only as a general guide. The expert collector and dealer also look at the color, physical nature, style, stamps, provenance, etc.

Lalique art glass
The Frenchman Rene Lalique produced some of the finest glassware. Lalique art glass from before 1945 typically fluoresces yellow and sometimes peach, but different colors after World War II.

Marbles
Many 1800s marbles fluoresce, some bright green and yellow. Many believe post WWII marbles do not fluoresce.

Vaseline glass
Vaseline glass is a popular form of yellow-green glass. Under black light genuine vintage and modern Vaseline glass fluoresces a bright green. Glass that resembles but is not Vaseline fluoresces differently, including peach, orange or less bright lime green.

Modern reproductions of Burmese Art Glass
Old Burmese art glass tends to fluoresces a bright yellow, while modern reproductions usually do not.

Dating American colorless pressed glass
American colorless pressed glass made from before 1925-30 fluoresces brightly, while modern reproduction usually do not.

Age of marble cut
Freshly cut marble tends to be a strong purple under black light, while marble cut a long time ago tends to be a mottled white. The marble surface gains a patina over time that changes the fluorescence. This test is good for checking if a marble statue or figure is old or new.

Cut jade
Jade that was cut just recently will be intense in color, while jade that cut a long time tends to be duller, mottled, patchy in color.

Carved ivory
Ivory that was carved recently tends to fluoresce purple under black light, while ivory carved a long time ago tends to be yellow, at least in areas, due to the patina gained with age.

(8)
Miscellaneous Uses for Your Black Light

This chapter looks at some practical and interesting uses for black light.

Invisible ink pen and secret messages

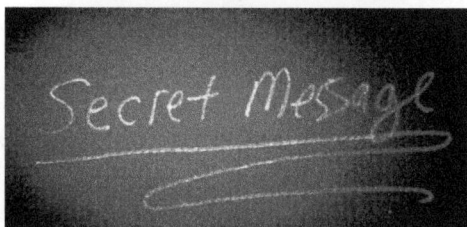

A useful, fun and inexpensive tool is the invisible ink pen. Looking similar to regular felt tip pens, invisible ink pens write with ink that is invisible in daylight but that fluoresces brightly under black light. UV ink pens are widely available online, including at amazon.com and eBay. Many sellers of black lights sell the pens.

People use invisible ink pens for a wide variety of purposes including writing secret notes to friends and making glow in the dark sketches.

Invisible ink is often used to secretly mark objects, including vases, paintings, prints and documents for later identification. A family might mark the bottom of valuable heirlooms in case of theft or dispute. You might write your name on the back of a painting to help later prove it is yours. If the marking was written in visible ink, a thief may scratch it off or mark over it. With invisible ink the same thief likely have no idea the painting is marked for identification.

A seller of radios parts or dolls may have a problem with customers who return for refund damaged goods they didn't purchase from the seller. If the seller puts an invisible ink mark on the back of the item before sale she can be sure that the customer is returning the item purchased from her.

UV ink hand stamps for dance clubs, bars and concerts
You can buy UV ink and ink pads for your rubber hand stamp. These can be used similarly to UV ink pens. Clubs often stamp the hand of patrons with invisible ink.

For authentication purposes, Major League Baseball put a UV fluorescent stamp on the ball Barry Bonds hit for his record 73rd home run in 2001.

Judging the authenticity of drivers and other licenses
Most state driver licenses have markings that are invisible except under black light. Credit and many other cards often have similar UV fluorescent markings. Black lights are used to authenticate licenses and credit cards. In the airport security line, look for the homeland security official who uses a black light to check your driver license.

Checking the cleanliness of kitchens and bathrooms
Under black light, invisible stains on sinks, tabletops, bathroom floors and the like often fluoresce. They often fluoresce yellow.

Black lights are used by health officials to check the cleanliness of motel and hotel rooms, public bathrooms and the like.

Finding rats and mice
As rat and mice urine fluoresces under black light, exterminators and biologists trace the rodents with a black light. You will often see the pest exterminators on television carrying a black light. Scorpions also fluoresce.

Making fluorescent art and crafts
With a black light and UV fluorescent materials you can make all sorts of glow in the dark arts, crafts and designs. This ranges from sketches to mobiles to sculptures to ceiling stars to Halloween decorations. Your imagination is your limit.

Invisible ink pen sketch on cardboard fluorescing in the dark.

Those inexpensive UV ink pens are great for drawing sketches and designs on paper, cardboard, wood and other surfaces. Though the UV ink does fluoresce on normal white paper, you may find it best to draw on material that does not fluoresce. This is particularly true if you want the sketch to stand out on its own, like a glowing skeleton in a dark room.

Of course the artist should be careful what he writes one. The ink may or may not be easy to remove or cover over. Writing a design on a disposable plastic toy a piece writing of paper is one thing. Writing on mom's favorite silk scarf is another. Also, it only makes sense that you don't want to write on areas that people eat or drink from. Use common sense.

You can make cut-and paste crafts using fluorescent materials like paper and minerals. You can cut fluorescent computer paper into designs, fold them into dolls and figures and hanging mobiles, use them to embellish your sketches, embellish them with invisible ink pens or stamps. On the market, there are fluorescent paint and inks that fluoresce in different colors.

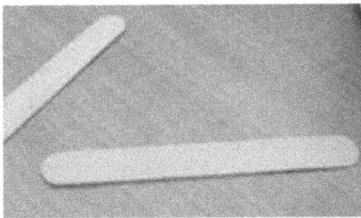

ghost tongue depressor

(9)
Shortwave/UVC Ultraviolet Light

Example of a light that gives off both longwave and shortwave light

Shortwave UV = UVC = Germicidal

Shortwave UV (UVC) is often used in combination with longwave UV (UVA, black light) when examining material. UVC is particularly useful with gem stones and minerals, postage stamps and some art glass. As with longwave, shortwave is invisible to human eyes and makes some materials fluoresce. With some material, the longwave and shortwave fluorescence will be identical. However, in other cases the fluorescence will be different, which helps in identification. The shortwave fluorescence may be a different color and/or intensity.

Shortwave is more dangerous than black light. This does not mean the adult reader can't or shouldn't use shortwave if needed for his area of collecting, but that following proper precautions are essential. The keys are to not look directly at the light, as it can damage the eyes, and to use the light as little as is needed. UVC is not good for the skin. A positive is that shortwave doesn't penetrate cloth or glass. For short term use, wearing glass goggles and long sleeved shirt and gloves will offer good protection. For normal uses in this book, you don't need to shine the light more than a few seconds every once in a while. It's where someone is using it for extended periods that it's a big concern, such as a scientist in a laboratory. Stamp collectors and rock hunters around the world use shortwave UV, so I'm telling you to be careful, not terrified. With proper and prudent usage, UVC can be safe.

Combined shortwave and longwave lights can be found on eBay and elsewhere on the internet, along with many science and rock shops. The light should allow the user to alternately use the lights, one light on at a time, as many tests involving comparing the brightness and color of longwave versus shortwave. Lights that are designed for gems and minerals will do this.

another type of long wave/shortwave UV light

UVC and postage stamp markings

Many modern postal stamps have special markers on them that can only be viewed under longwave and shortwave ultraviolet light. Some also phosphoresce under shortwave light. These markers are used for the automatic machine sorting and handling of letters, and are useful in identifying counterfeits.

As discussed in an earlier chapter, the backlight can also identify many modern counterfeits of antique stamps, as the modern paper will fluoresce very brightly due to the presence of optical brighteners.

Canadian stamp and fluorescent mark under UVC

Shortwave/longwave fluorescent minerals

Identifying raw minerals in the field is a complex task, requiring lots of knowledge and experience. However, double checking if that stone on that ring really is a ruby as advertised is more straightforward. There are numerous methods involved in identifying minerals and gemstones, including checking the fluorescence under both longwave and shortwave ultraviolet light. Other methods include looking at the general appearance, shape, visual color, relative hardness (for example diamond is hardest so scratches all other gems and glass). The following are a few examples of how ultraviolet light aids in authentication. Ultraviolet fluorescence doesn't in and of itself authenticate a mineral or gem, but it is a useful check.

* Natural black pearls fluoresce a dim red, while dyed versions do not fluoresce.

* Natural yellow sapphires fluoresce yellow, synthetic yellow ones do not fluoresce. Natural colorless sapphires fluoresce orange, synthetic colorless do not fluoresce. Synthetic orange sapphires fluoresce red.

Shortwave UV in Irradiation, Purification and Sterilization

UVC light, typically at 254 nanometers, is used to irradiate, purify and sterilize water, air, food, sewage, laboratories, office buildings, ponds and aquariums. Industrially the process is called Ultraviolet Germicidal Irradiation (UVGI).

Direct and long enough exposure to UVC light can kill, amongst other things, anthrax, typhoid, e. coli, tuberculosis, salmonella, dysentery, strep, bird flu, cholera, flu, hepatitis, algae, and fungi. The UVC breaks down the molecular bonds of the organisms, making them unable to reproduce.

There are limitations to UVC as a germicide. There must be direct exposure for it to work. This means in a lab or building there may be areas that are missed when a germicidal light is shown. Because of this, laboratories often used UVC in combination with other sanitary methods, such as applying bleach in the nooks and crannies.

UVC is commonly used to purify circulating air, as in an office building or medical clinic. While the air is purified at the spot of exposure, the air can get dirty again as it circulates.

There are commercially marketed hand held UVC germicidal lights for the public. They are commonly used by travelers to sanitize hotel and airplane bathrooms, beds, dishes, etc. I've even seen a UVC toothbrush cleaner.

While one should take due care using UVC light, there are advantages of UVC over other traditional disinfectant and purification methods. UVC doesn't harm the ozone, there are no dangerous fumes as with bleach and ammonia, and there is no liquid runoff to contaminate lakes and rivers.

UVA offers about no practical purification or cleaning effect. UVB offers some. As regular sunlight includes UVB, it does have some disinfecting effects, including due to the drying.

Did You Know?

Danish Physician Niels Finsen (1860-1904) won the 1903 Nobel Prize for Medicine for studies of how ultraviolet light treats diseases, including Tuberculosis.

(10)
Forensic Visible Light

As visible light is common to us and measurable by our own naked eyes, forensic visible light is far from an esoteric subject. When a baseball card collector examines a card with a loupe or naked eyes, he is using visual light. When a shopper holds a shirt up to a store light to look for tears or snags she is inspecting using visible light. Eyes are our most common instrument for detecting and measuring visible light. Even in the ultraviolet test, like seeing how materials visibly fluoresce under UV, we are using both UV and visible light. We observe the visible light produced by ultraviolet light.

This chapter looks at several visual light techniques commonly used in examination. These and other visible techniques are applied in other chapters, including authenticating currency, examining paper and trading cards.

Opacity—the 'see through' effect

Opacity is measured by the amount of light that shines through a material, or the 'see through' effect. The object is held up to a good light source, like a desk lamp. The object will let none, some or lots of light through. The brightness of the light source has to be taken into consideration when assessing opacity as it affects how much light goes through.

Opacity is a good way to identify paper, cardboard, plastics and other materials that differ from each other. Opacity is often used to directly compare cardboard objects from mass produced issues, like trading cards. The opacity is often consistent through a single trading card set. A questioned card's distinctly different opacity can help identify it as a reprint or counterfeit.

Holding an object up to a bright light is also useful for identifying restoration, cracks, tape residue, overpainting and such.

Different brands of white paper held up to the sun. The left sheet clearly lets through less light than the right sheet. This type of difference can help identify a forged letter or reprinted art print.

Gloss

Identifying the gloss (or lack thereof) is often helpful when judging an item. An object's surface can be glossy (reflects lots of light), matte (reflect little to none) and in between. In many cases, an object will have different glosses on different sides. A vacation postcard may have a different gloss on front versus back. Many trends are commonly knowledge and can be applied even without having before examined known genuine items. For example, fellow collectors might tell you when you start collecting that 1950s Bowman baseball cards are glossier on the front than back and that many reprints will have the same gloss on both sides. Many counterfeit trading cards are first suspected due to the atypical gloss. "If the 1933 Goudeys aren't supposed to be glossy," a newbie collector may ask, "why is mine so glossy?" This will cause the newbie to look further into the issue, perhaps showing it to a more experienced collector or sending it into a grader. The gloss isn't proof in and of itself of a counterfeit, but is evidence leading to further investigation. If a variety of other qualities are bad (size, thickness, coloring, other), the collector will be more and more sure he has a fake.

Many photographic processes are identified by the gloss, or lack thereof, on the front and back of the photo. This is important for dating and authentication, as certain processes and gloss qualities existed for a finite period of time. Chromogenic is the name for the standard color photo, including the old family snapshots and 8x10s celebrities autograph. Vintage chromogenic color photos are glossy on front and matte on back, while recent chromogenic color photos are glossy on front and very smooth on back. A matte back chromogenic photo of James Dean (died 1955) is consistent with the photo being original, while a smooth back shows that the photo is a modern reprint. The Polaroid has a glossy image and back, but a matte border surrounding the photographic image. As Polaroids were introduced in 1963, a Polaroid of James Dean is easily identified as a modern reprint. Popular in the early 1900s, the cyanotype photograph is identified due to the bright blue image (cyan = blue) and that the paper is matte front and back. The modern cibachrome color photograph is identified in part as it is ultra-glossy front and smooth back.

Silvering on photo as sign of old age. Many, though not all, early 1900s black and white photos have a distinct quality called silvering, or silver mirroring, that appears in the dark areas of the image. It looks as if silver has risen to the surface. This silvering is revealed as the photograph is viewed at different angles to the light, the silvering changing from light to dark at different angles. As silvering is an aging process, happening over many years, it is near definitive proof that a photograph is old. Collectors and historians will often look for silvering on an expensive antique photo to make sure it is antique.

Shining light at different angles

Shining light at various angles at an object's surface helps reveal qualities and identify clues. Identifying an object's gloss, or lack thereof, involves looking at the surface under different angles of light. Dust, cracks, added materials (paint, glue, varnish, other), texture are often identified when the surface is viewed at a near parallel angel to the light.

Magnification
Whether using a loupe or high powered microscope, magnification involves bending light from its normal path to create an enlarged view. This is helpful in getting a closer look, identifying wear and small wrinkles and, in the case of microscopes, getting a look not remotely possible with the naked eyes.

Example of Magnification: Prints Authentication
Over the centuries there have many different types of printing processes: etching, engraving, digital, lithography, woodcut, etc. Being able to identify and know about the printing types is an essential part of authenticating everything from movie posters to Rembrandt etchings, vintage trading cards to magazine covers. Many reprints and counterfeits are identified because the printing is the wrong type or from the wrong era to be original. For example, a lithograph can't be an original Rembrandt, as lithography was invented after Rembrandt's death. 1960s Fleer brand trading cards are lithographs, so a photocopy is identified as a reprint or counterfeit.

Microscopic examination is one of the keys to identifying and dating printing, as it allows the expert to see key details invisible to the naked eyes. To give you a taste for this type of examination, the following are 100 times magnified images of three different kind of printing.

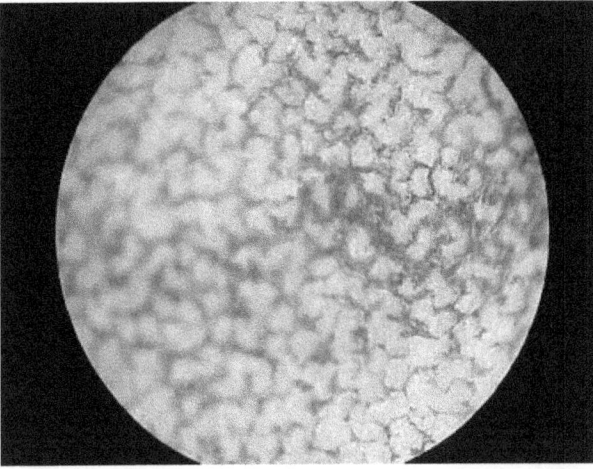

This shows the macaroni-like ink pattern of a collograph. Collagraphs were used to make many early 1900s postcards and commercial prints. Few to no recent reprints and counterfeits will have this printing pattern. This pattern doesn't in and of itself authenticate a 1912 picture postcard, but is a strong sign that the postcard is antique.

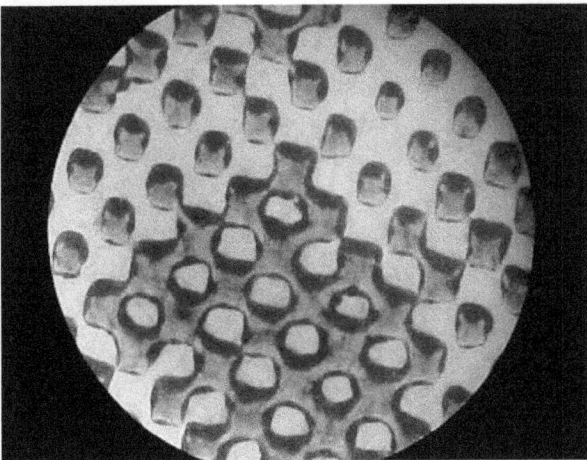

This shows a detail of an old time commercial printing process, called photoengraving. The 'waffle' pattern with dark rim around the edges helps identify. If an advertisement, poster or trading card has this ink pattern, that's a strong sign it is vintage.

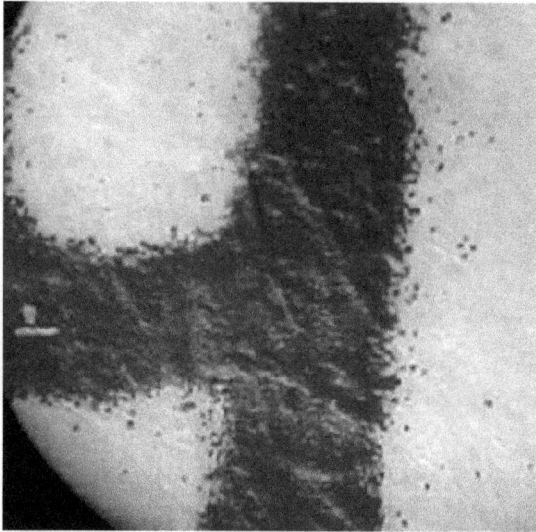

The stray dust-like specks identify this as Xerography, the printing used for Xeroxes, photocopiers and laser printers. Many of our home computer printers are laser printers. Very few fine art, commercial or other collected prints are made with xerography. If a trading card, currency, ad sign, original Picasso or antique print is a xerograph, its easily identified as a cheap fake.

Using a light meter

For quantification and documentation of gloss and opacity, you can buy a relatively inexpensive tool called a light meter. These give a numerical reading of light measured in lux. Though it takes practice, trial and error and patience, these objective readings are important information and research. You can objectively measure the opacity a trading card and the gloss of a vase.

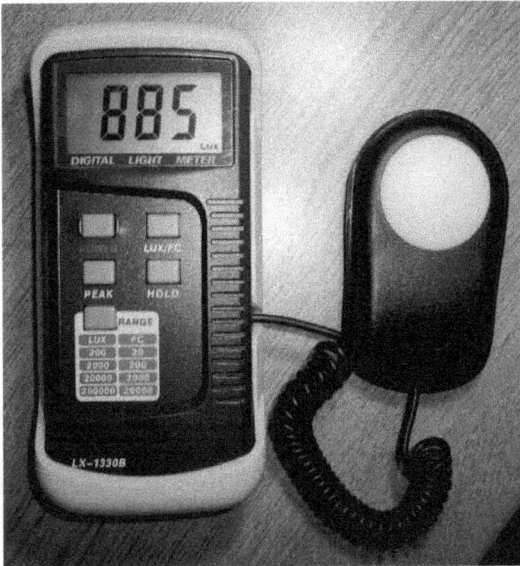

Example of a light meter. The left part gives the reading, while the right part receives the radiated or reflected light.

(11)
Infrared Light

**Infrared camera : A regular digital camera can be
converted into an infrared camera and viewer.**

The infrared camera and viewer (you only need one or the other,
take a pick) allow one to view objects in the infrared range.
Unlike ultraviolet light, infrared doesn't make things visibly
fluoresce. Rather, the camera or viewer translates into visible
light the infrared light that naturally reflects or radiates off of an
object. You see the visible translation in the form of an image
on the screen of an infrared camera or viewer. With an infrared

camera, you can see this image plus take a digital photograph of it. Infrared cameras and viewers are easy to use.

As with ultraviolet light, the infrared viewer allows us to see details and qualities from a different point of view than normal. Things that may appear dark in visible light can be light through the infrared camera, and visa versa. As in all areas of life, more points of view are important to making informed judgments.

Infrared viewers and cameras are safe to use, even safer than back lights. Unlike the ultraviolet lights, the infrared camera and viewer do not emit light but view light that already exists

Infrared view: Some things look different in infrared light. Tree tops, grass, bushes, leaves and flowers (above) usually appear snow white in infrared light. Clothes, yarn and blankets also often appear white.

As with ultraviolet light, infrared is commonly divided by wavelength into categories. The following are common categories:

IR-A (Near Infrared): 700 nm–1400 nm
IR-B: 1400 nm–3000 nm
IR-C: 3000 nm–1 mm

<u>The range used in this book is the IR-A or Near Infrared range, roughly 700 to 1000 nm.</u> This is the infrared closest to visible light. Most digital infrared cameras and infrared viewers you find for sale are in this range.

Common style of hand held infrared viewer allows viewing in the IR-A or Near Infrared range

Should you buy an infrared camera or viewer?

The infrared camera and viewer are the most expensive pieces if equipment described in this book. A viewer can cost hundreds to even thousands of dollars, while a converted digital camera can be purchased for a few hundred bucks. I own an infrared digital camera, a converted Fuji brand, and am happy with it. It's easy to use, shows and photographs quality infrared images. It handles like a normal digital camera, with the images downloading onto my computer. I also own a much more expensive infrared viewer, and like the digital camera better. The camera's images are of better quality and the viewer doesn't take photos. Obviously, if a collector were to ask me, I'd recommend getting the camera over the viewer, both due to cost and usability.

Considering that even the camera is expensive, the big question is will the reader get enough use out of the camera to justify the cost? If an infrared viewer or camera was as cheap as a black light, I'd say everyone should get one, heck even two. However, due to cost, it's up to reader is decide if it's required for his purposes. For the average baseball card, stamp or poster collector, an infrared viewer is not required. For someone who wants to do advanced forensic studies, including of paintings and cloth, the infrared camera may be a good investment.

Infrared art photography

A bonus with the infrared camera is that with it you can take interesting art photos of nature, people, places. Infrared is a part of modern art photography, with the images giving a dreamlike view of things. In other words, when considering cost, there is an additional use for your infrared camera beyond forensics. Most infrared digital cameras are made, bought and sold with art photography in mind, not forensics. The infrared world is as easy to photograph as the visible world— point, focus and shoot.

Purchasing an infrared viewer or infrared camera

Infrared digital cameras are the normal digital cameras that many of us have and use, but have been converted to view only or mostly infrared light. Inside a digital camera is a filter that blocks out infrared light. In the conversion to an infrared camera, that filter is replaced with a filter that blocks visible light and allows infrared light. You can find already converted infrared cameras on eBay and elsewhere. Make sure it comes with the normal software and directions that allow you do down load your images onto your computer.

Infrared Reflectography: Seeing Through Paint

In the historical art and artifact world, infrared viewers are best known for their ability to view through the top layer of paint of a painting. Art historians and museum conservators view through the paint to see any background sketches or earlier versions of the art. They do this to learn about how the painting was made and how the artist worked and planned and changed things. Studying the style and types of changes and background images is useful in determining if a painting was by a famous painter— his habits and techniques already being known. This process is referred to as infrared reflectography. Museums also use X-rays to do similar examinations, but X-ray technology is out of the reach of most readers. Infrared and X-ray examine different levels of the paint.

How well an infrared camera sees through paint depends on the color, type and thickness of the paint. Typically, the infrared camera allows one to see through some to most of the top layer of paint, but not all. Thin, light colored paint usually allows the clearest view to paint behind it. Dark, thick paint often prevents light penetration.

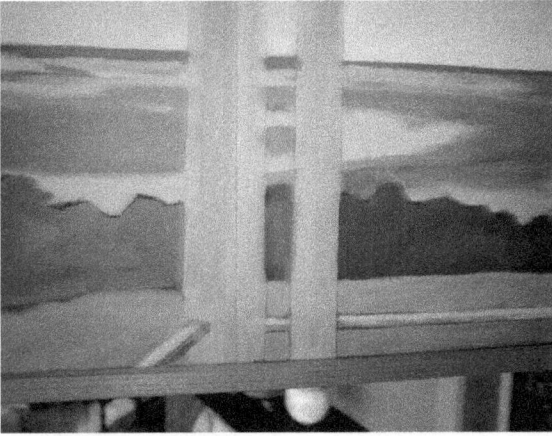

The above shows visible light detail of a modern acrylic painting. The below shows a closer detail photographed with an infrared camera. You can see the bushes other details previously hidden by the posts. This shows how the artist made the painting: first bushes, then posts painted over the top.

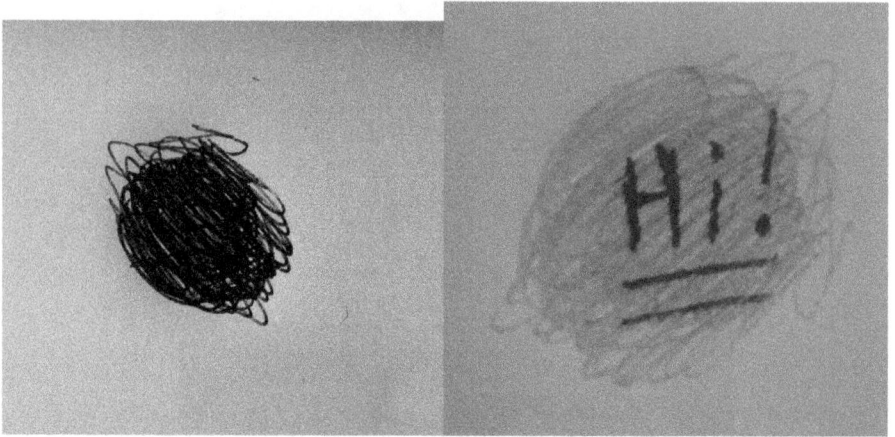

Hidden message: On the left is seem a ballpoint pen scribble in visible light. On the right is an infrared camera photo of the same scribble, showing a secret message in pencil. The infrared camera sees through the top ink layer to the pencil writing below. This not only shows how an infrared camera can see though ink and paint, but how one can make secret security markers on valuables and documents. On a valuable object, document or heirloom, the security code or owner's name can be hidden beneath paint, ink or a sticker.

An infrared viewer can help in identifying some items that have been repainted or touched up. The touched up area may be unnoticeable in visible light, but may stand out in the infrared range/ Also the viewer may be able to see through the current layer of paint on a to see old designs underneath. If a table was originally painted with a flower patters then repainted a light color, the original pattern may be viewed beneath.

Studying and examining the trends and making comparisons of materials can be helpful in identifying fakes and reprints. A modern reprint may look different than the original in the infrared range. This use of infrared is limited compared to black light, but still helpful. Again, it allows the examiner or collector to view material in a different light. The reader should examine materials and note trends.

Infrared photograph of house and trees

(12)
Identification of
Restoration and Alterations

Forensic light is helpful in identifying many types of restoration and alteration to paintings, prints, furniture, photos, vases and more. These items can be altered by the addition of paper, glue, paint, varnish and/or other material. Items are typically restored to fix damage and make things appear new.

As the added material often fluoresces differently than the rest of the item, the restoration can often be identified under black light. Similarly, an infrared viewer will often reveal differences in material.

To identify alterations, one should also look for visual differences in texture, gloss and opacity. When an art print is put at an angle nearing 180 degrees to a light, the added paint, ink or paper will often have a different texture and gloss from the rest of the surface. The added material may be physically raised from the rest of the surface. You might be able to feel a relief with your finger tip.

The backs and insides of items often reveal restoration—for examples, the back of a pin may reveal solder or the inside of a desk drawer may reveal the original varnish.

As restoration and alterations affect financial value, the presence of known alterations must be revealed at sale. A mint condition movie poster is worth more than the movie poster restored to look mint condition. Neatly trimming and touching up a baseball card will drastically lower the value, even if it looks like new. Not disclosing known alterations at sale is unethical and often fraudulent.

Some dealers and collectors remove autographs from baseballs for aesthetic or financial reasons. For example, a single signed Joe DiMaggio baseball can be worth more than the same ball with the bat boy's signature beneath. There is one or more companies that will remove autographs. While the removal may be difficult to see under normal daylight, the restoration usually easily shows up under black light.

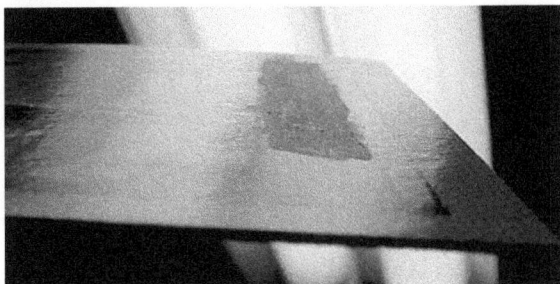

Holding this baseball card at an angle to a disk lamp, added paint is clearly seen due to the difference in gloss. This paint can also be felt with the fingertips.

In some cases forgeries are alterations. For example an inexpensive baseball card may be changed into a rare and valuable error or other variation by changing text. In the earlier mentioned driver license forgeries, the forgery may be a genuine license that has had the date of birth altered. In many cases these changes are identified by the just mention techniques. In a few cases the forger covered the entire card or license in a clear varnish-like substance to try and cover up the handiwork. The substance however gives the item a different gloss and black light fluorescence than normal. This is detected by comparison to other cards or licenses. A collector didn't notice the altered text of one baseball card, but noticed the card had a distinctly different gloss than his other cards from the same set. Closer examination by an expert revealed the alteration.

As with ultraviolet, infrared light can be useful in identifying restoration and alterations, as the different materials can contrast with each other, and as infrared can sometimes see through the first layer of paint, ink or paper. Touched up areas might stand out in the infrared range. A totally repainted antique toy may have a different infrared tone than original versions.

Did You Know?

How infrared and ultraviolet were discovered

Infrared was discovered in 1800 by the famous German astronomer William Herschel. Herschel was studying the colors of light produced by a prism. Using a thermometer as part of the experiment, he was surprised to discover that the area beyond the color red was hotter than in the visible spectrum. He concluded that there must be an invisible form of light, which we now know as infrared. Infrared translates to 'beyond red.'

Ultraviolet light was discovered the following year by scientist Johan Wilhelm Ritter. After hearing that Hershel discovered a form of light beyond the color red, Rittner experimented to see if invisible light also existed beyond the color violet at the other side of the visible spectrum. He discovered that silver chloride turned black under ultraviolet light. Silver chloride is found in many old photographs as it turns dark under sunlight. Using a prism, he spread apart light into its rainbow spectrum, and saw that an invisible light beyond the violet turned the silver chloride black. This showed that an invisible form of light existed beyond the violet end of the spectrum. Ultraviolet is Latin for 'beyond violet.'

(13)
Paper

When judging the authenticity of a possibly centuries old print, being able to identify the type of paper is essential

Having a basic knowledge of paper is important in many areas of collecting. Many letters, certificates, autographs, books and fine art prints are identified as fakes because the paper is too modern or the wrong type. Visual techniques and black light are commonly use to inspect paper. As you read in an earlier chapter, black light can identify much modern paper, fakes and reprints.

This chapter is a more in depth look at the different types and qualities of paper.

The following are standard types of paper.

Laid paper: Until the 1750s, all paper was laid paper. It was made on a mesh of wires about an inch apart, with finer wires laid close together across them. This gridiron pattern can be seen when the paper is held to the light. Today, some writing tablet paper is still laid, though the pattern being more of a decoration.

A paper print, sketch or letter from the 1500s or 1600s has to be on laid paper. An original Rembrandt etching or engraving couldn't be on wove paper, as he died in 1669.

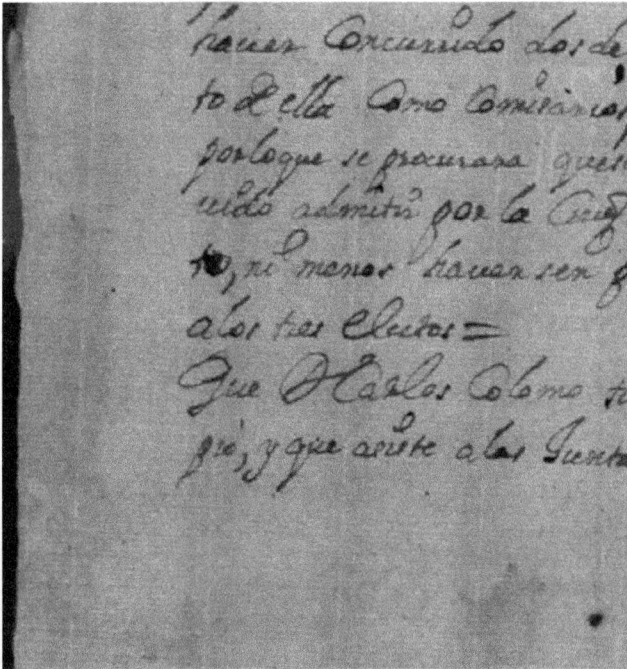

This centuries old letter shows the gridiron pattern of laid paper.

Wove paper: About 1755 wove paper was invented. Wove paper is made on a finely woven mesh, so the paper does not have the rigid lines pattern of laid paper. Laid and wove paper are easily differentiated when held up to the light. Most of today's paper, including computer printer, is wove. No print from before 1750 could be on wove paper. Many reprints of centuries old prints are identified due to the wove paper.

Rag versus wood pulp. In the early history paper was made from rags. Starting about the mid 1800s, rag pulp began to be replaced by wood pulp. Wood became a popular choice due to the scarcity of rags and because wood pulp paper was cheaper to manufacture. The first successfully made American wood pulp paper was manufactured in Buffalo New York in 1855. By 1860, a large percentage of the total paper produced in the U.S. was still rag paper. Most of the newspapers printed in the U.S. during the Civil War period survived because they were essentially acid-free 100% rag paper, but the newspapers printed in the late 1880s turn brown because of the high acid content of the wood pulp paper. In 1882, the sulfite wood pulp process, which is still in use today, was developed on a commercial scale and most of the high acid content paper was used thereafter in newspapers, magazines and books.

Counterintuitively, modern paper, especially in books, letters and newspapers, is much more likely to turn brown and brittle than paper from before the American Civil War. The paper on an early 1800s print can be surprisingly fresh and white, while a 1950s newspaper is often brown and decrepit.

* * * *

Chronology of Paper

The following is a brief chronology of paper history. Paper has been traced to about 105 AD China. It reached Central Asia by 751 and Baghdad by 793, and by the 14th century there were paper mills in several parts of Europe.

> 105: Paper making invented in China.
> 610: Papermaking introduced to Japan from China.

868: Earliest printed book, the Diamond Sutra, in China.
900: First use of paper in Egypt.
1228: First use of paper in Germany.
1282: Watermarks first used in Europe.
1319: Earliest use of paper money in Japan.
1450-55 Johan Gutenberg's forty two line Bible produced.
1470: First paper poster, in the form of a bookseller's advertisement.
1521: First use of rice straw in Chinese paper.
1589-91 European printing introduced to China and Japan.
1609: First newspaper with regular dates (Germany)
1662: First English newspaper introduced
1750: Cloth backed papers introduced. Used for maps, charts, etc.
1755: Wove paper introduced
1758: First forgery of bank notes
1824: First machine for pasting sheets of paper together is introduced. Cardboard is first formed.
1830: Sandpaper introduced commercially.
1830s: Coated paper introduced. This paper is usually coated with China clay, which makes it white and smooth, sometimes glossy. It is most often used in art and illustrated books.
1842: Christmas card invented.
1844:First commercial paper boxes made in America.
1862:Tracing paper introduced commercially
1871: Roll toilet paper introduced.
1875: First instance in U. S. of paper coated on both sides.
1903: Corrugated cardboard introduced. Replaced many wooden boxes.
1905: Glassine paper introduced
1906: Paper milk-bottles introduced
1909: Kraft paper introduced
1910: Bread and fruit wrapped in printed paper

* * * *

Some common fine art paper terms. For collectors, dealers and authenticators of art, these terms often come up and it is important to know what they indicate.

Blind stamp: an embossed seal used to identify the artist, publisher, printer or collector.

China Paper: a soft paper made in China from bamboo fiber.

Chine appliqué, or chine collé: A chine appliqué is a print in which the image is pressed into a thin sheet of China paper which is backed by a thicker and stronger paper. Some proof prints are chine appliqués.

Cold pressed: A paper with slight surface texture made by pressing the finished paper between cold cylinders.

Deckle edge: the rough, almost feathery edge on hand made paper.

Deckle Stain: Paper that has a coloring or darkness around the deckle edge.

Drystamp: blind stamp.

Embossment: A physically raised or depressed design in the paper.

Enameled paper: any coated paper.

Glassine paper: A super smooth, semi-transparent paper that is often used to make the envelopes that hold stamps

Hand made Paper: Paper made by hand in individual sheets.

Hot Pressed: A paper surface that is smooth. Made by pressing a finished paper sheet through hot cylinders.

India paper: an extremely thin paper used, primarily in long books to reduce the bulk.

Machine Made Paper: Made on a machine called a "Fourdrinier." Produces consistent shape and textured paper.

Marbling: a decorative technique of making patterns on paper

Mouldmade Paper: paper that simulates hand made paper, but is made by a machine.

Parchment: An ancient form of paper made out of animal skin. It is smooth and semi-translucent.

Plate Finish: A smooth surface made by a calendar machine.

Rag Paper: Made from non-wood fibers, including rags, cotton linters, cotton or linen pulp.

Rough: a heavily textured paper surface

Tooth: A slight surface texture.

Vellum: a modern version of parchment, with the same dense, animal skin-like appearance. A slightly rough surface and is semi-translucent.

Some drafting paper is called vellum.

Velox: Black and white paper print for proofing or display

* * * *

Watermarks

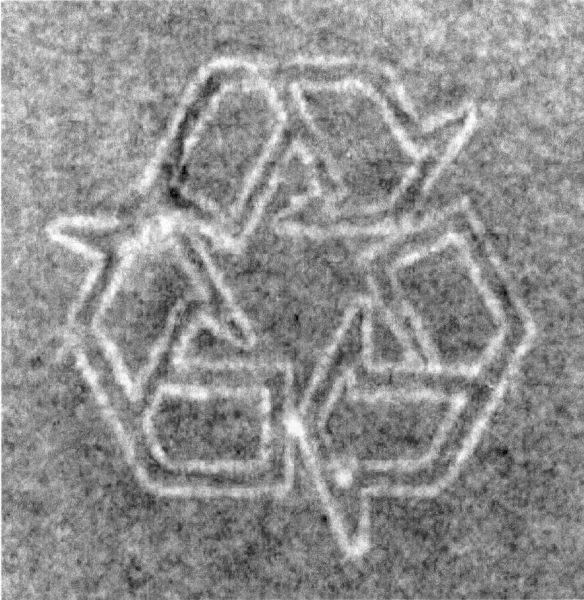

A watermark in paper is best seen when held up to a light

For centuries paper manufacturers have made watermarks. A watermark is a design in paper made by creating a variation in the paper thickness. The watermark is visible when the paper is held up to a light. Watermarks can sometimes give important information about the age of the paper and the authenticity of the print or other document. Watermarks have been made since at least as early as the 13th century.

Two types of watermark have been produced. The more common type, which produces a translucent (lighter) design when held up to a light. The rarer *shaded* watermark, which is darker when held up the light, is made by making the paper denser in the area of the watermark design. Watermarks are often used commercially to identify the manufacturer or the grade of paper. They have also been used to help detect and prevent counterfeiting and forgery. A bank might put its watermark on official financial documents to help both the bank, other banks and the customers identify forgeries.

Examples of how watermarks help identify art prints:
If a Salvador Dali print has a watermark consisting of the word "ARCHES" with an infinity sign (sideways '8') beneath, the print is a fake. Dali used ARCHES brand paper, but in 1980 ARCHES added the infinity sign to the watermark. 1980 was past Dali's working career and Dali himself stated that he never used the 'infinity' paper. While this watermark is easily identified, some enterprising forgers and dealers, picked the 'infinity' paper where the watermark was near an edge so they could conveniently cut off the infinity. A simple rule of thumb for collectors is to make sure that you buy a Dali print on Aches paper where the watermark is entirely on the paper and away from an edge. Thus you see whether or not the watermark has the tell tale infinity sign.

For John James Audubon's large size *Birds of America* prints, the presence of a "J. Whatman" watermark is strong evidence that the print is original. No known reprints or later restrikes are on paper with that watermark.

Pablo Picasso sometimes used paper with his personal watermark.

Letters of authenticity from noted autograph experts PSA/DNA and James Spence Authentication have company watermarks. Letters missing their watermarks should be assumed to be not original.

Rarely, but sometimes, watermarks are forged by drawing in linseed oil. The linseed oil makes the paper more transparent. This forgery is usually to identify, as it the linseed oil usually fluoresces under black light. A sniff test might help as well, as linseed is smelly.

(14)
Identifying Counterfeit US Currency

There are numerous methods used for identifying counterfeit US currency bills, including the use of black light, visible and infrared light. Note that this section is only a brief and general introduction. Currency is regularly changed and updated by the US Government, and it's likely that there will be new changes to the design and details within a few years of the publication of this guide. However, newly issued currency doesn't make old currency disappear. Old currency is floating around for many years. If you find a 1930 $1 bill it is valid currency.

Counterfeits vary in quality, from easily identifiable to sophisticated. The following techniques will help identify most counterfeits, even many of the toughest ones.

The following looks at specific techniques of counterfeit detection. An important thing to realize is that a single correct quality does not prove a bill authentic. For example, some counterfeiters bleach genuine $1 bills and reprint on the paper to make fake $20 bills. That the paper itself is genuine doesn't prove these fakes authentic. However, there more than likely will be other qualities that do identify them as fake.

Pay attention to your currency
Observe your bills before you get counterfeits. Look at the printing, size, the Presidential portrait, examine the details, get a feel for the paper. A common way a counterfeit bill is found suspect is that it looks and feels off, different from other bills. The image may look funny and unclear, the color may be off, the paper may feel too stiff or too soft.

Genuine bills have high quality, detailed printing. Check the details and lines in the portrait and in the background. The detail in reprints is often lacking and muddled.

Compare a suspect bill to known genuine bills of same denomination and year. Again compare the feel and general look. Compare specific, close up details, like the President's eye or the points on a seal. Compare all the designs and text. Again, remember that the design and text changed over the years on genuine bills, so you want to compare bills from the same year.

Take into consideration that there can be natural differences between genuine bills. One genuine bill can be crisp and unused, while another genuine bill can be worn, wrinkled and dirty. This is why comparing to numerous bills is a good idea.

Black light test #1: fluorescent vertical bands
Some recent currency above the $1 denomination have vertical bands that fluoresce different colors under black light. Under normal visible light, the bands can be seen when the bill is held up to a light. The presence of these is strong evidence of authenticity.

The bands fluoresce the following colors under black light:

$100 Pink/Orange
$50 Yellow
$20 Green
$10 Red
$5 Blue

UV fluorescent bands on a $5 and $20. Notice they are in different spots. This helps identify counterfeits made from bleached real bills, such as a $5 bleached and reprinted as a $20.

Black light test #2:
Authentic currency does not have optical brighteners in the paper. Many, but not all, counterfeits are made with paper with optical brighteners and will fluoresce brightly.

Infrared camera/viewer test:
Some recently issued higher currency, including the US$20 and $5 bill have a horizontal band on the left side of the back that can only be seen under an infrared viewer. The band will appear as a

blank strip amongst the normal printing. This test is very reliable in identifying authentic paper.

Bills viewed under an infrared camera, revealing the blank bands. There are different band patterns for each denomination.

Watermarks. Modern higher currency bills have a watermark to the side of the bill. The authentic watermark is not seen until it is held up to a light. It will be a smaller portrait of the president on the bill and can be seen when viewing from both sides.

2006 US$ bill with watermark. The left is the normal, everyday view. On the right, the bill was held up to sunlight revealing a watermark. If $5 bills are bleached and printed over to make higher denomination ($20, $50, other), the '5' watermark will identify them as fakes. Many bills, including this one, have watermarks of the President on the bill. Again this will help identify bleached counterfeits. An Abe Lincoln watermark shouldn't appear on an Andrew Jackson bill.

Fibers in paper. Some modern currencies have real thread-like fibers of different colors in the paper. In some cases you can remove the ones on the edge with a pin.

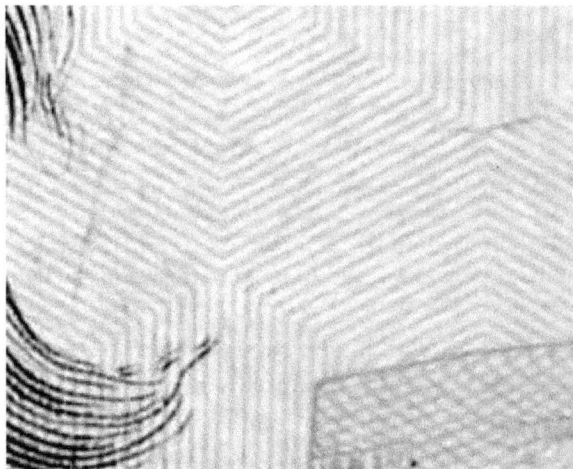

colored fibers can be seen in the paper

Microprinting: Microprinting is very, very small text that appears in some parts of some but not all currency. It is readable under magnification and very hard to reproduce in a counterfeit. In most counterfeits, the microprinting will be blurred or missing under magnification.

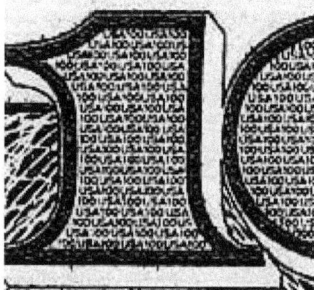

Microprinting: even under this magnification, you still can't clearly see that it says *USA100* over and over inside the numbers.

Color shifting ink on higher than $5 currency: On modern higher currency, there is a distinct color shifting ink used on the front right. It has a metallicy finish and is used on two right side symbols. It changes color from green to black when you change the angle of the bill. This is hard to duplicate in counterfeits.

Minute multi-color dot pattern as identifier of counterfeits: When you examine a genuine bill under good magnifier, you will see the images, text and design are comprised of solid monotone lines and marks. Many cheap counterfeits are identified by a minute multi-color dot pattern in the printing. The dot pattern is not visible to the naked eye. Even the blank borders and other blank areas can have the dot pattern. Most digital, computer, photocopy and Xerox counterfeits have this dot pattern.

Raised notes. Some genuine notes are altered to give them a higher denomination. For example a forger may take a $1 bill and paste '$10' on the corners. This forgery is easily identified by knowing which presidents appear on which bills. George Washington appears only on a $1. Also, the correct denomination is spelled out just below the President's portrait.

Paper testing pen. There are inexpensive commercially available pens that test the currency paper. Genuine currency is fiber based, while many counterfeits are on wood pulp paper. Your computer paper is wood pulp based. The pen contains iodine that makes a black stain on the wood pulp paper, but not on fiber-based. The black stain shows that the bill is counterfeit. Realize that some counterfeits are made on fiber-based paper, including bleached genuine currency, so the pen won't identify all counterfeits. Many foreign currencies are also on fiber based paper, so the pen will work with the Euro, Mexican Peso, Indian Rupee, Greek Drachma, German Mark, French Franc, British Pound, Russian Ruble, Japanese Yen and numerous other paper currencies.

Did you know?

While humans can't see ultraviolet or infrared, some animals can. Snakes can detect infrared light, while bees, geese and butterflies can see longwave ultraviolet.

(15)
Judging the Authenticity of Trading Cards:
Comparing to Known Genuine Cards

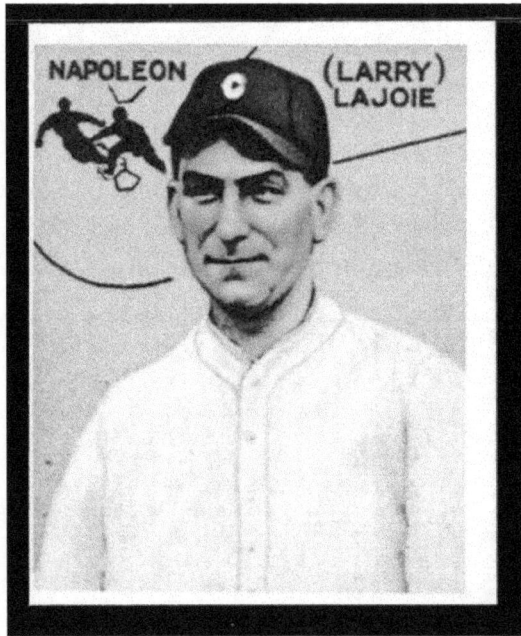

Cheap reprint of the expensive and rare 1933 Goudey Napoleon Lajoie baseball card. The original in strong condition is worth tens of thousands of dollars, this reprint probably less than $1.

A standard and often highly effective way to detect counterfeits and reprints of trading cards is by directly comparing the card in question with one or more known genuine examples. This includes comparing the fluorescence and visible light qualities, like gloss and texture. Granted, it is uncommon for the collector to already own duplicates, especially of an expensive card like a

1933 Goudey Babe Ruth or 1965 Topps Joe Namath. However, good judgment is often made when comparing a card to different cards from the same issue. Comparing the Ruth to a bunch of Goudey commons and the Namath to a handful of other 1965 Topps. This type of testing is also useful with other mass produced items like stamps, comic books, magazines and postcards.

An expensive 1909 T206 Ty Cobb baseball card, and even a million dollar T206 Honus Wagner, was printed on the same sheet as T206 common cards. The printers did not bring out special cardstock and VIP inks for the superstars. When you are studying the qualities of cheap T206 commons, you are also studying the qualities of the T206 Honus Wagner.

If there are cards insufficient in number or of extra poor quality (caught in the back yard thresher), other techniques outside the scope of this book will be required for authentication.

In nearly all cases, counterfeits and reprints are significantly different than the real card in one and usually more than one way. However, in other cases, even though a difference or two is identified (cardboard a bit thinner and lighter in color), this doesn't answer whether the difference is due to fakery or is a genuine variation. Over a long printing run, some variations are to be expected.

Before examination, the collector should be aware of variations within an issue. A genuine 1956 Topps baseball card can be found on dark grey or light grey cardboard. While the 1887 Old Judge tobacco cards are usually sepia, pink examples can be found. The examiner must also take into consideration reasonable variations due to aging and wear. A stained card may be darker than others. An extremely worn or trimmed card may be shorter and lighter in weight than others in the issue. A card that has glue on back will allow less light through when put up to the light. The collector will often have to make a judgment call when taking these variations into effect. This is why having experience with a variety of cards is important.

The following is a short list of things to look at.

Obvious Differences: This can include text or copyright date indicating the card is a reprint (1995 copyright on a reprint of a 1933 card), major size difference, wrong back text. Many of these problems are obvious even in an online scan.

If you are experienced with an issue, perhaps you've collected 1930s Goudeys for the last few years, most reprints and counterfeits within that issue will be obvious. They simply will look bad. The experienced eye is a sophisticated tool.

Solid areas: With a magnifier or microscope, compare which areas are solid and which are not. On a genuine T206, the white border around the player picture and the player's name and team below is solid. While many reprints will also have these areas solid, many will not. They will often have a dot pattern under magnification.

On the 1971 Topps baseball cards (example pictured on page 74), the front has a printed faux player autograph. On the genuine card this signature is in solid black ink. On a home made computer reprint, the faux signature will almost always be made up of a multi color dot pattern.

Appearance of card stock and surfaces: This includes color, texture, feel, etc. The correct gloss is hard to duplicate on a reprint, and most reprints will have different gloss than the original. Make sure to check both sides. A T206 and 1951 Bowman, for example, have different textures front versus back. Make sure to check the thickness, color and appearance of the card's thickness or edge. The edge often shows the cardstock to be different. Comparing the cards' weights is useful. A Counterfeit usually has a different weight than the genuine card.

**The reprint on the right has a distinctly different gloss
and coloring than the original card.**

Font and size of lettering and border lines: Some reprinters go
to the effort of recreating the lettering and border lines, making
them solid like with the originals. In many of these reprints, the
font of the lettering is noticeably different than on the originals.
This includes the thinness of the lines, height of the letters, and
the distance between lines of lettering. If you are familiar with an
issue, the lettering on one of these reprints will be strikingly
different on first glace. Similarly, the border lines and designs
may be noticeably different. In a few cases, the counterfeiter left
out entire words from the text.

Opacity: Opacity is measured by the amount of light that shines through an item, or the 'see through' effect.

Cardstock and ink vary in opacity. Some allows much light through, some allow none, while the rest will fall somewhere in between. Most dark cardboard will let through little if any light. White stocks will usually let through more. While two cardboard samples may look identical in color, texture and thickness, they may have different opacity. This could be because they were made in different plants, at a different time and/or were made from different substances. Testing opacity is a good way to compare card stock and ink. The same cards should have the same or similar opacity.

Opacity tests should be done with more than one card from the issue. Comparisons should take into consideration variations due to age, staining, soiling and other wear, along with known card stock variations in the issue. It must be taken into consideration that normal differences in ink on the card will affect opacity. If one genuine T206 card has a darker picture (a dark uniformed player against dark background), it should let less light through than a genuine T206 card with a lighter picture (a white uniformed player against a light sky).

The opacity test can detect many restored expensive cards. Some genuine but low grade star cards (1933 Goudey Ruth, T206 Cobb, etc) have been restored in part by having the rounded corners rebuilt with paper fibers from other cards and glue. When held to the light, the built up corners are often seen as they let through a different amount of light than the rest of the card.

Comparing Opacity: When held to a normal desk lamp, the 1971 Topps Hank Aaron reprint lets through much more light than the original 1971 Topps Tom Kelley.

Black Light Test

Studying the degree and color of fluorescence under a black light is an unbeatable tool for comparing ink and cardboard. If you spread out in the dark a pile of 1983 Topps baseball cards with the exception that one is a 1983 OPC (Canadian-made version of Topps cards), the OPC will be easy to pick out with black light. The OPC is made out of a different card stock and fluoresces many times brighter than the Topps stock. The OPCs have other distinct qualities that help separate it from Topps, including visible color of the card stock and roughness of the edges. Topps have sharp, smooth edges, while OPCs have fuzzy edges not unlike the edge of a rug.

This is the way it often works for reprints and counterfeits. Reprints and counterfeits were made with different cardstock and often fluoresce different than the genuine cards. The reprint may fluoresce darker, lighter or with a different color. In some cases, a reprint and an original may fluoresce the same, but in most cases the black light will pick out the reprints with ease.

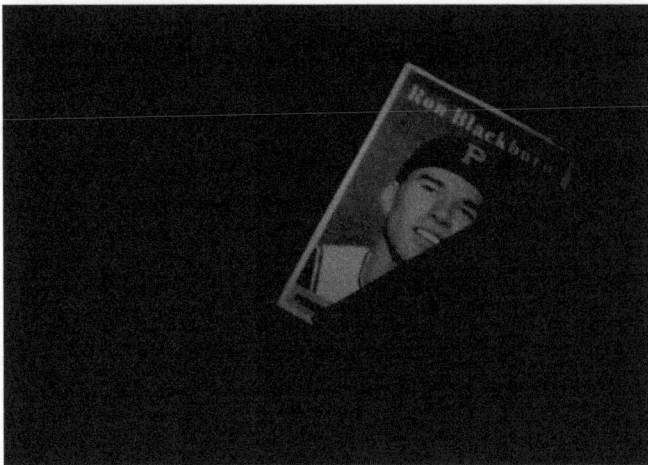

The top picture shows six original 1958 Topps baseball cards and one reprint. Under black light, it's obvious which one is the reprint.

* * * *

Sometimes, the differences between a questioned card and genuine examples will be significant enough that the collector will be nearly certain it is a fake. If that 1984 Topps Dan Marino rookie card has a significantly different gloss, thickness, fluorescence and opacity from genuine commons in the issue, the card is almost certainly a reprint. In other cases, reason for the difference won't be so clear, and a second opinion may be required. For example, there may be a different color tone, but you can't be sure if it's to reprinting or a genuine printing variation.

Many collectors buy and sell cards that have been graded and authenticated by reputable grading companies. The below listed three are widely considered reputable. Cards deemed authentic are put in transparent plastic holders with labels on top. The label will include the grade, Poor to Mint. Realize that there are dubious, fly-by-night grading companies. Just because a company calls itself a grader doesn't mean its judgments are trustworthy.

Reputable third party trading card graders
PSA (Professional Sports Authenticators) psacard.com
SGC (Sportscard Guarantee) sgccard.com
Beckett Grading Services   beckett.com

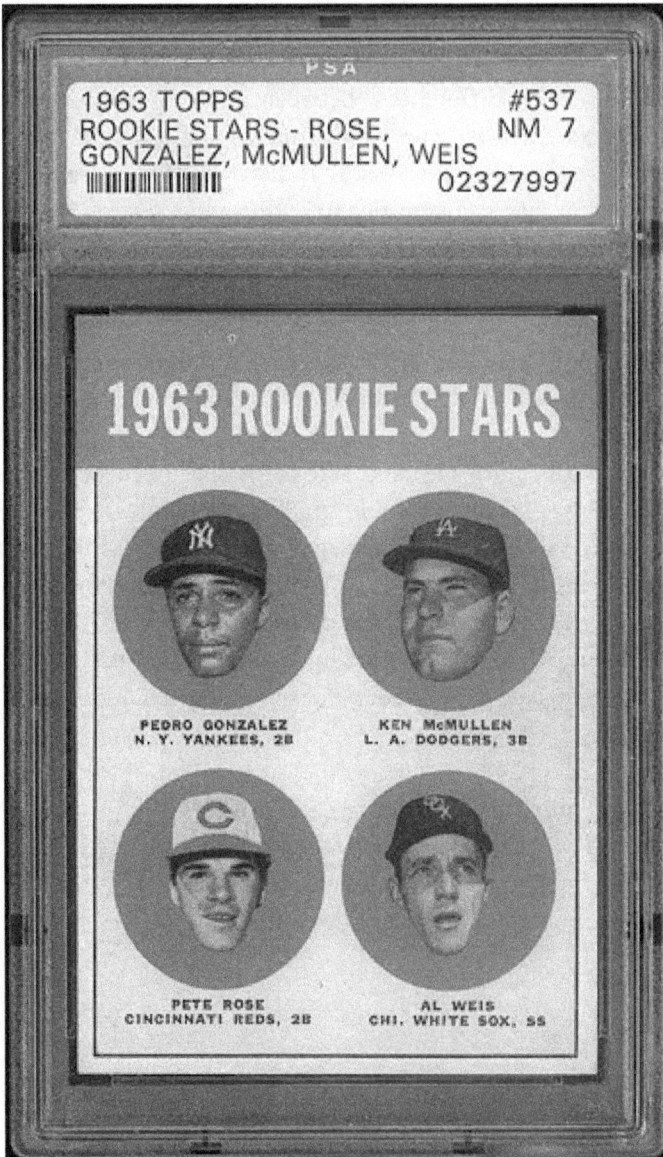

1963 baseball card graded by PSA. The card is held in a
holder with detailed label at top. This card happens to be
the rookie (first year) of baseball great Pete Rose.

(16)
Examining Cloth

Many expensive collectibles are cloth or part cloth. This
includes a baseball cap worn by a famous player, a rare
American Civil War uniform and antique teddy bears. Some
expert collectors and examiners of valuable clothes and cloth use
forensic light to judge age and identify fakes and alterations.
Examples of fakes include a modern reproduction sold as an
antique original, and a game worn baseball jersey where the
name plate and number has been changed to that of a more
valuable player.

Visual light, especially opacity, is good for identifying
alterations, restoration and wear and for comparing cloth.
Holding a garment up to the light is an age old method of
looking for tears and alterations.

Black light is also good for identifying trends in cloth. It is
great at identifying new cloth with optical brighteners. As with
paper, optical brighteners have been added to many clothes made
after World War II. Used to make bright colors brighter and
stain resistant, the optical brightened clothes will fluoresce a
bright white or blue/white under black light. The optical
brighteners will typically indicate that the cloth was made after
World War II. Many faked antique patches, hats and shirts are
identified as being modern reproductions-- or at least recently
repaired or altered-- due to the presence of the presence of
optical brighteners. For example, collectors of WWII military
patches know that it's a recent reproduction if a patch fluoresces
brightly under black light. A person who bought an antique style
New York Yankees baseball cap will be able to identify it as a

modern reproduction by tags and stitching that fluoresce brightly. Thread can fluoresce very brightly and modern thread itself can identify a modern reproduction or recently altered antique.

As many cloth items are made from a variety of cloths and threads, the optical brighteners will often appear only on parts. For example, most of a modern baseball cap might not fluoresce, except for the emblem, stitching and laundry tag. Cloth or thread that doesn't fluoresce brightly doesn't mean it isn't modern. In fact, most dark modern cloth doesn't contain optical brighteners.

It is important to note that many laundry detergents have optical brighteners that can throw off results. If an antique shirt was washed in the washer, it may have optical brightener residue from the detergent. The granular detergent is usually fairly easy to identify as detergent due to the granular, dusty pattern.

Cloth tends to lose its UV fluorescence with time, and very old cloth often has no fluorescence.

Examiners and collectors identify alterations on new and old cloth by looking for clear differences in UV fluorescence. If a coat is patched up with a like color of cloth, the alteration can be identified by a UV fluorescence difference between the patch and the rest of the coat. This type of comparison judgment requires experience.

An infrared camera offers an additional view of the cloth to supplement the black light and visual views. It can sometimes identify different cloths. On very old textiles, alterations often stand out in the infrared range.

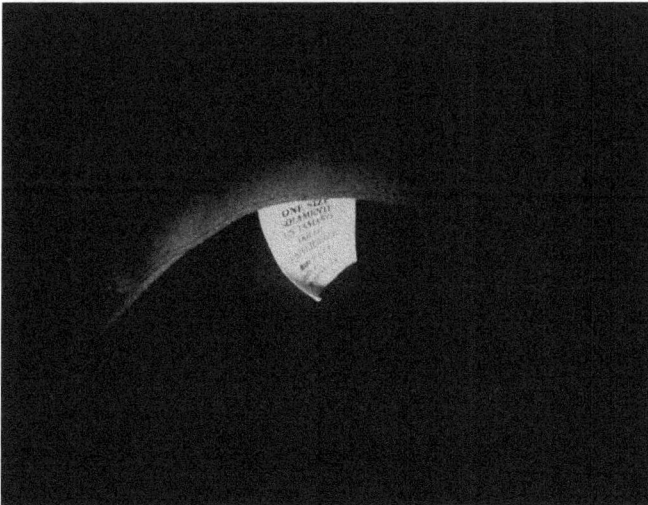

**The brightly UV fluorescing tag (below picture)
reveal the old looking hat to be modern.**

(17)
Crime Scene Investigation

Detectives and forensic scientists use ultraviolet, infrared and visible light in their examinations of crime scenes and crime scene objects. They use the same light described in this book, but usually employ a more expensive machine that gives off many more gradations along the light span. The light used is often much brighter than we use in this book. Different visible color lights are used, including blue, yellow, green and orange. In conjunction with these colors, the scientists wear color tinted glasses. The right combination of glasses color and light color can make items stand out, like fingerprints.

Ultraviolet light is used to find bodily fluids that fluoresce, including saliva and semen. Blood stains don't fluoresce but shining a range of light can make it contrast against the background. As some backgrounds fluoresce and make the object of interest harder to see and photograph, changing the light wavelength can create a better contrast. The expert will go through the full ranges of light to see what is uncovered at the scene. Going through the full range of light can help identify faint bruises, bite marks and foot prints. The key is to experiment until you find the light that produces good contrast with the surrounding background.

Photographing the scene and objects is an important part of crime scene forensics. Finding a good contrast to make objects and qualities stand out is doubly important for this. The photos may be used as courtroom evidence.

Special UV fluorescent powders are used to find and photograph finger prints, with the powders sticking to the prints. As antifreeze fluoresces, black light is used to determine the path and speed of cars in car crashes. As much modern cloth fluoresces brightly do to optical brighteners, black light is used to help find lost bodies submerged in lakes and rivers. Advanced forensic scientists use different forms of UV and IR to compare and identify pen inks in the laboratory.

Bright light at different angles is used to find fine, small material. Hairs can sometimes be found with close naked eye examination or with magnifying glass. Hair sometimes fluoresces under ultraviolet light. Bright visible light shined parallel to a surface (floor, table, bedspread) will often identify hard to find hairs. The visual light used is usually much brighter and with a fuller spectrum than on a standard flashlight.

84

About the author

David Rudd Cycleback is an art historian specializing in the issues of authenticity and cognition. He is photography advisor to Beckett Media, runs the Applied Light Laboratory at cycleback.com, has advised and examined material for major auction houses and was a contributing writer for Encyclopedia of 19th Century Photography. He has been cited by Australian National Archives, Indiana Historical Society, Encyclopedia Britannica, University of Wisconsin's *The Scout Project*, Main Antiques Digest, PBS and Sydney University's *The Business of Art*, and his books have been required reading in university courses internationally. His other books include *Judging the Authenticity of Prints by the Masters*, *Judging the Authenticity of Photographs* and *A Look at How Humans Think and See*.

cycleback.com